EL SUPER

Kurt Hollander / khollander@laneta.apc.org

Photographs
Adam Wiseman / adamphoto@prodigy.net.mx

Graphic Design
Rocío Mireles, Bruno Contreras / despachomireles@gmail.com

First edition, 2006

US Distribution   **D.A.P.**
Distributed Art Publishers, Inc.
155 Sixth Avenue 2nd Floor
New York NY 10013
www.artbook.com

Other countries   **Books on the move Actar S.L.**
Actar D
Roca i Batlle 2-4
08023 Barcelona - SPAIN
office@actar-d.com
www.actar-d.com

ISBN: 968-5208-47-6   Editorial RM, S.A. de C.V.

All rights reserved.
No part of this publication may be reproduced, stored in a retrieval system or transmitted, in any form or by any means, electronic, mechanical, photocopying, recording or otherwise, without the prior permission of Editorial RM.

Printed in China

EL SUPER

KURT HOLLANDER

**WITH PHOTOGRAPHS BY:
ADAM WISEMAN**

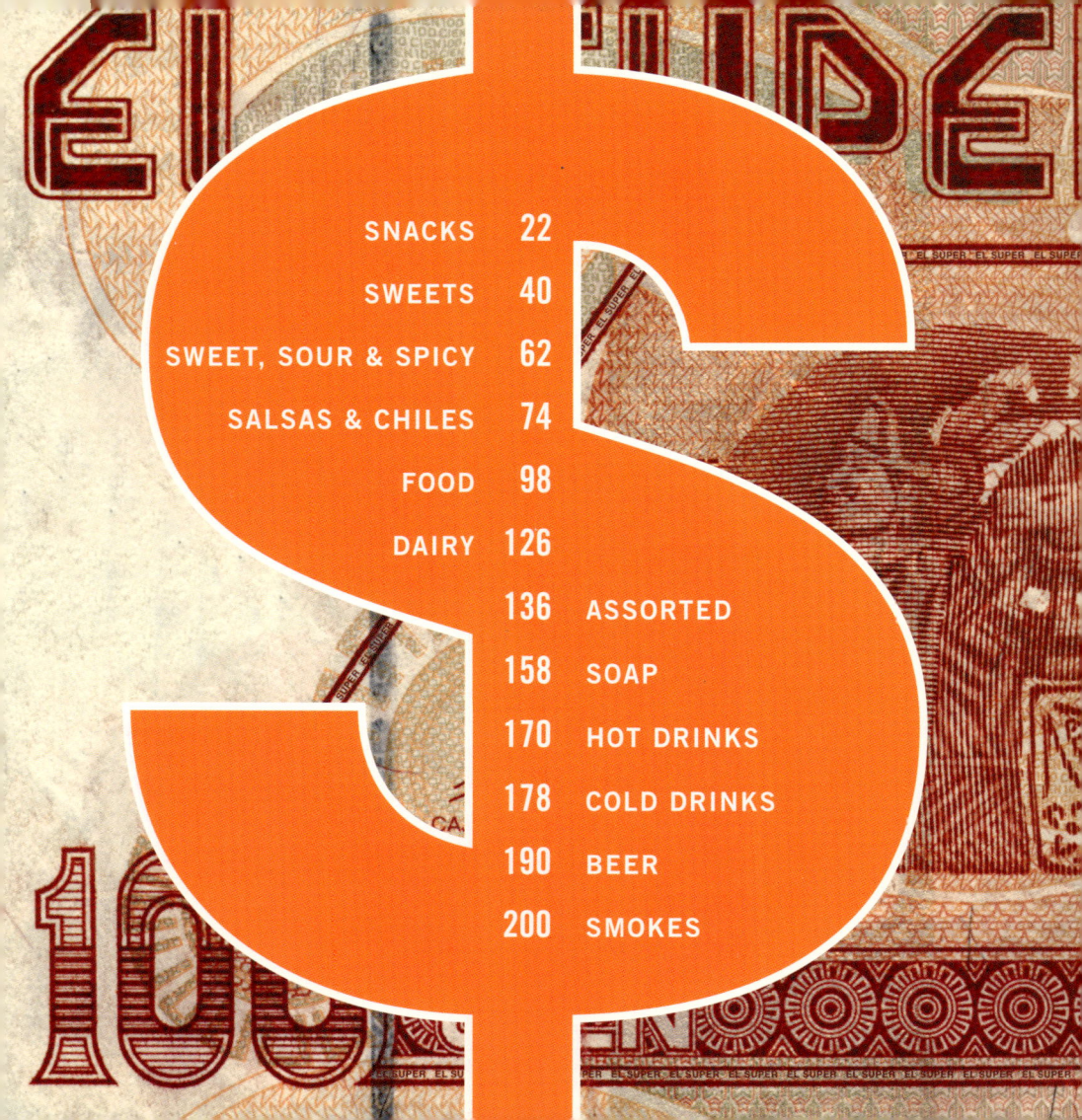

SNACKS	22		
SWEETS	40		
SWEET, SOUR & SPICY	62		
SALSAS & CHILES	74		
FOOD	98		
DAIRY	126		
		136	ASSORTED
		158	SOAP
		170	HOT DRINKS
		178	COLD DRINKS
		190	BEER
		200	SMOKES

NEZAHUALCOYOTL
PESOS

IN MEXICO, *EL SUPER* MEANS THE SUPERMARKET,
although it also means to go shopping, buy groceries (as in *hacer el super*). Most of the products included in this book were purchased in supermarkets in Mexico City, such as the up-scale Superama owned and operated by Wal-Mart, the working-class Gigante and Sumesa, or the government-subsidized ISSSTE. Other products come from the Central de Abastos, the largest distribution market in Latin America; neighborhood *mercados sobre ruedas* (markets on wheels); OXXO, 7-11 (known colloquially as *Super 7*) and other retail chain stores; small corner stores and street vendors.

 If it is true that you are what you eat, then shopping would be the way you construct your own personality, and the way a culture constructs its own identity. The products that appear in this book were designed for local consumption, not for export to other countries and cultures, and thus they do not present an exotic, foreigner-friendly view of Mexico. The packaging of these products has come to represent images of what a majority of Mexicans have in mind when they think about their own contemporary culture. El Super is an attempt to collect these images that together form a cultural identity.

DESIGN HAS ALWAYS PLAYED A CENTRAL ROLE IN MEXICO.
The Aztecs were a visually exuberant culture. The ceremonial feathers, body painting and piercings, the comic-like pictograms and murals, as well as the jade, silver, and gold ceremonial knives and jewelry, give witness to an advanced aesthetic and a very boastful palette. In turn, colonial Spain left its mark on the nation's visual culture, with its stylized script, its religious icons painted thick with blood and gold, its gaudy clothes and rococo architecture. The Mexican muralists brought overblown scaling and glossy colors to the walls of Mexico, while José Guadalupe Posada introduced cartoon-like skeletons and gore to the pages of news print.  For the Aztecs, and for Mexican culture since, all that glitters is good. Hence the gobs of make-up, the masses of dyed blondes, the omnipresent shoeshine stalls, the flashy polyester clothes, the gold chains and watches, the shiny imported luxury cars, and the cheap glitz of national TV shows and pop singers. Contemporary consumer goods sold in stores around the country reflect these aesthetics, and their design reflects this wealth of tradition.

The term *naco*, a word probably derived from Totonaco, an indigenous civilization, is used in contemporary Mexico by the upper classes to describe gaudy, tacky products, the people who buy, drink or wear them, and lower-class culture in general. *Naco* exists in popular visual and commercial expressions in public spaces in Mexico: the posters for *lucha libre* or *salsa* bands; the hand-painted signs and neons; and in the design of food, beverage and household products sold in stores.

Naco is the Mexican equivalent of white trash in the United States, a culture that elicits the same response from the cultured classes. *Naco* is the cultural opposite of the bourgeois notion of good taste, good taste being very important for the local elite whose self-image depends on how their lives compare to that of the Europeans and other cultured cosmopolitans. In fact, however, upon closer inspection *naco* turns out to be everything that is authentically Mexican, those home-grown products that are best adapted to the palettes and tastes of its people.

Naco, however, is not the aesthetic that the upper classes want to be associated with Mexico. Mexico, unlike the United States and Europe, is still a mostly working-class country, and middle-class values have not yet dominated all cultural and social representations. Vendors in the *metro* still sell books about Aztec myths, local comics and lurid pulp magazines still sell millions of copies more than any glossy magazine, and quesadillas and other deep-fried delights sold on the street still outsell Domino's pizza and McDonalds burgers. Canned goods still occupy a large section of most stores, and the contents are sometimes so "exotic" that refined tastes would be scandalized by the uncouth ingedients, such as tripe stew, or nose and cheek pig meat. The class struggle is between people of good taste and people who like products that taste good.

THE TRADITIONAL MEXICAN DIET, AS IT HAS EXISTED

for thousands of years, based around corn tortillas and beans, is a healthy diet, one that has sustained millions of economically struggling people. The health problems that Mexicans are currently suffering from (diabetes, obesity, heart problems, cancer) come from changes in this diet, changes that are relatively recent. Like the developed countries who have suffered from these problems for decades, **the contemporary diet for the majority of Mexicans is now rich in the four basic food groups; sugar, fat, alcohol, nicotine.** With the important addition of chile.

Modern-day industrial processes and chemicals used to produce greater quantities of products in less time to reach more consumers also cause health problems. Processed foods have had a devastating impact not only on the health of Mexicans, but on several essential aspects of their life. The devastation of the Mexican countryside in the 60s and 70s, which led to urban overpopulation and wide-spread poverty, came about in large part due to the transformations induced by the industrialization of agriculture, the very industry that began to produce processed foods.

These non-traditional food stuffs, such as processed wheat-based products, canned goods, salty snacks and fat-saturated foods, have radically changed the millennial diet of Mexicans, while sweetened beverages have created junk-food junkies of the latest generations. Consumers who have whole-heartedly embraced this new diet, pumping up on surplus fat and carbohydrates, have seen their bodies sculpted in new dimensions with added curves.

Being that Mexico is and has for centuries been a major sugar producer, there are naturally thousands of variations of sweets sold everywhere at very affordable prices. There are also countless products that combine chile and sugar, often with tropical fruits. The bitter tamarind, which no one eats as is, is a favorite children and adult snack when covered with sugar and chile, sweet and spicy and sour, something that makes most refined palettes cringe.

Sugar, like other stimulants such as tobacco, decreases appetite and gives a surge of energy to its consumer (well-adapted to those living in poverty). But like other addictive substances it also depletes the system of energy and health in the long run. The shift from sugar cane (and honey) to processed white sugar has changed a rich foodstuff into a health hazard, while the shift from cane sugar to imported beet sugars and artificial sweeteners has devastated the national sugar industry.

It is interesting how the products that one really needs to survive, those that provide the basic elements of a healthy diet, such as meat, fish, poultry, vegetables, are all their own advertisement, their flesh the packaging's only design. On the  other hand, junk-food products, stimulants and condiments that are in no way essential to survival and in no way add any nutritional value, are those that display the greatest amount of creativity in packaging.

IN MEXICO, SUPERMARKETS ACCOUNT FOR ABOUT

one quarter of the retail food market, while much of the remaining sales originate in small "mom-and-pop" stores and street vendors scattered across the country. Although these smaller, neighborhood outlets continue to represent the bulk of sales for many products, things are changing rapidly, especially in the major cities, and the shelf life of many Mexican products is dwindling.

The mega-chains of US mega-stores, such as Costco and Price Club, along with the Mexican-owned supermarkets, are luring the Mexican consumer away from neighborhood markets, some of which have been around for over 500 years, and away from regional and traditional products. The same phenomenon is occurring with the shift from the neighborhood stores to 7-11 and other national and global chains.

After the signing of the North American Free Trade Agreement, the competition from huge multinational corporations has reached historic levels. Brands like Hershey's and Marlboro, for example, with their aggressive marketing campaigns and promotions, are moving into traditional Mexican products (chocolate and tobacco are indigenous to Mexico) and gobbling up increasingly large slices of the national market.

The design of product labels and brands is also undergoing transformations, as the more artisan nature of Mexican traditional design now has to compete with design in the age of digital media. In addition, several of the large companies in Mexico are contracting design firms from the First World to give their products a more international style. National labels and brands, however, still often use pre-Colombian images to sell their products to a mass market comprised of the descendants of the Aztecs and Mayas. Mexican product design also has a fondness for images of other cultures, such as that of Europe, Africa, and even Asia. Nowadays, however, specific cultural images are being displaced by global characters from Disney films and cartoons, which are literally changing the face of food and beverages in Mexico.

Yet Mexican industry still dominates Mexican culture, and Mexican products still service the needs of the vast majority of Mexicans. Although many of these products are sold in packaging that is less-than-environmentally friendly, display images that are neither politically nor gender correct, and use ingredients that are banned in developed countries, Mexican consumers have nonetheless adopted these products as their own and see themselves reflected in their labels and designs.

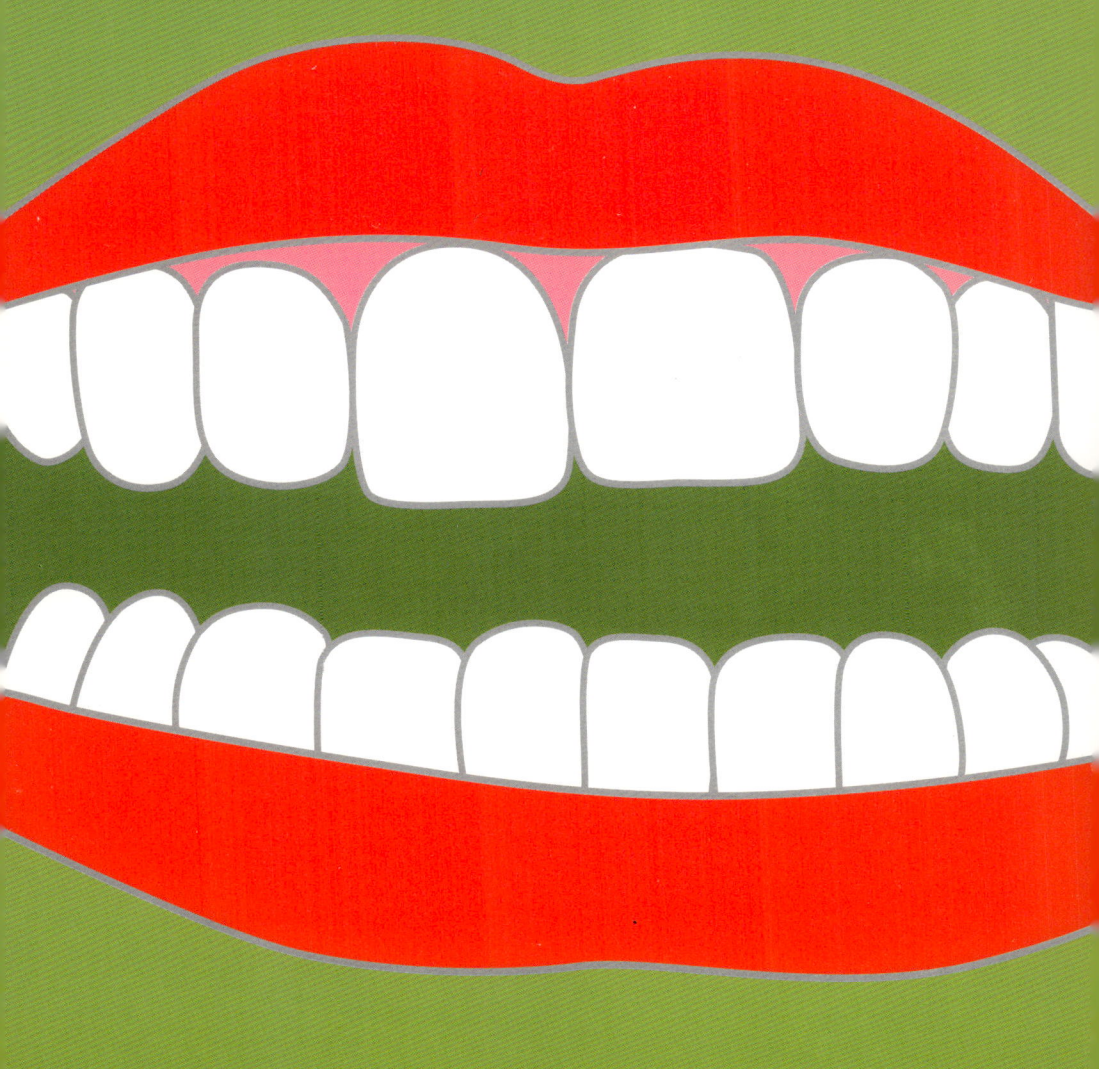

SNACKS

La Malintzi Amaranth bars

Oyuki · Corn and wheat sticks

Nishikawa — Japanese peanuts

Balmoro 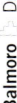 Dried apples

Papiux · Potato chips

Nipon Fried wheat squares

OvarB Fried pork rind (for snacks and stews)

ARINA DE TRIGO CON CHILE, S

YUK

Oyuki — Fried wheat chips

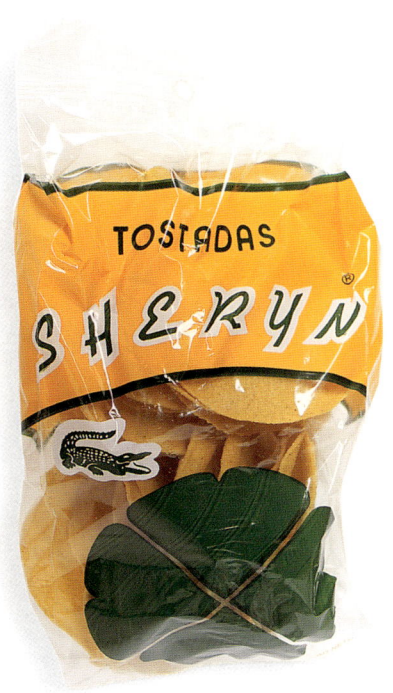

SWEETS

Canel's — Chewing gum

Las Sevillanas — Goat-milk caramel wafer

De La Rosa | Peanut bar

Borrachitos △ Milk caramels with wine

Cocada  Coconut bar

Selz Soda Candy

Manita de la Suerte — Lollipop

Oblea Mini Cometa — Wafers

Fresta, Pecsi, Fanda 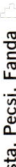 Candy

SWEET, SOUR & SPICY

Gusano Lucas — Tamarind with chile

Tiramindo Tamarind with chile

Miguelito ▷ Salty, spicy and sour sugar

Miguelito Salty, sweet and spicy pulp

Muecas — Tamarind with chile

Pelon Pelo Rico — Tamarind with chile

SALSAS & CHILES

Chimay | Hot sauce

El Yucateco   Hot sauce

Cholula ▷ Hot sauce

Herdez Hot sauce

Herdez Sun-dried chiles

Valentina · Hot sauce

Mydac | Dried chiles

La Consentida | Hot sauce

Tajín — Powdered chile

Salsa Huichol — Hot sauce

FOOD

La Merced Dried shrimp

El Trébol Chicken tamales

Tamal de Elote — Corn tamales

El Titi ▸ Corn niblets

Pullin — Dried beef

Tía Lencha — Dried beef

Zwancito — Spicy pig-liver pate

San Juan — Egg

Nopalmex

Nopal
PRECOCIDO

Obleas, las sevillanas — Obleas con cajeta de leche de cabra

CALIDAD INTERNACIONAL

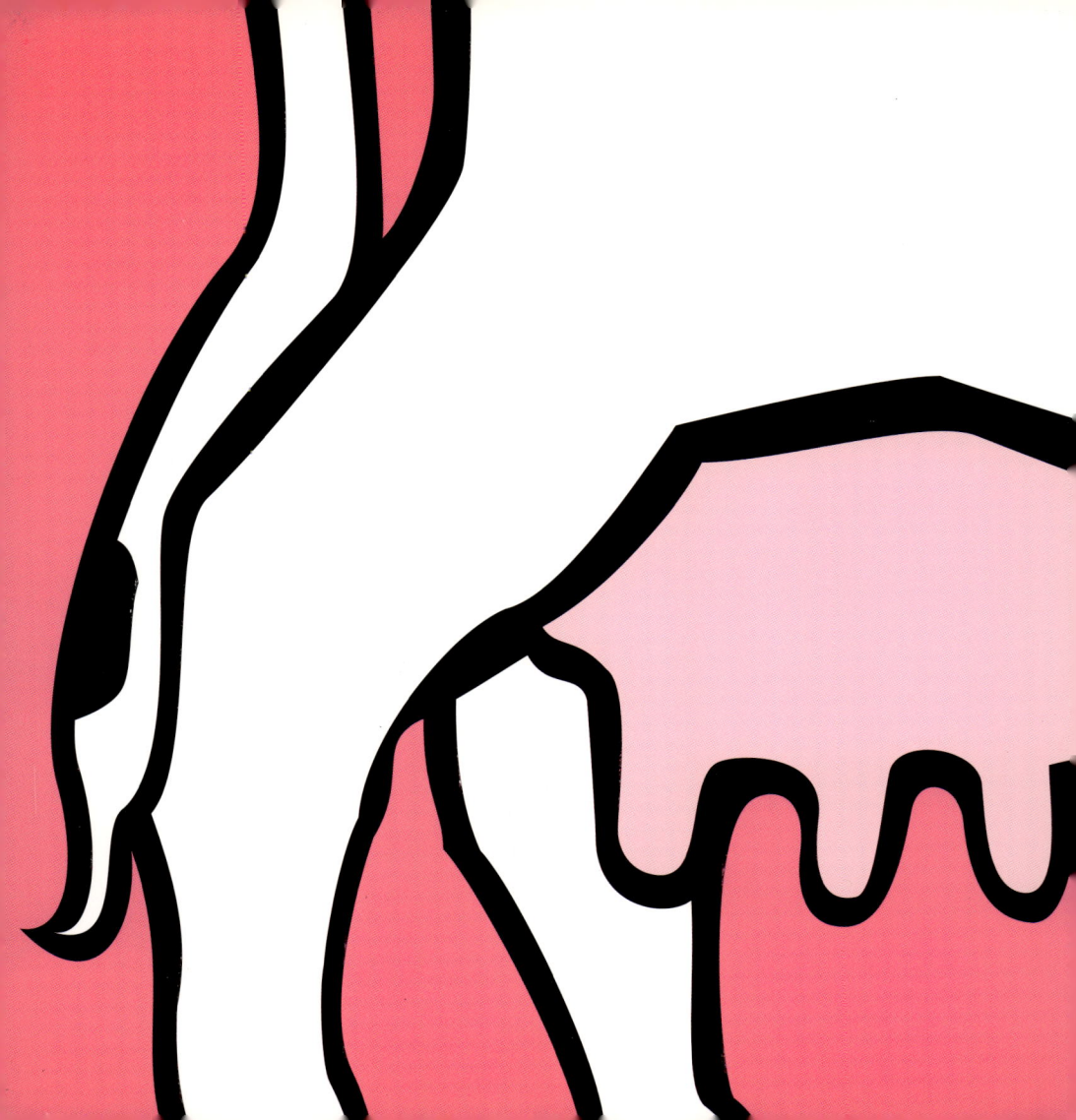

La Abuelita — Sour cream

Chipilo ▷ Butter

Gloria — Butter

Yakult ▷ Fermented lactose product

INGREDIENTES: ACEITE DE GIRASOL / CANOLA 0,007% ANTIOXIDANTE TBHQ

EITE **COMESTIBLE VEGE**

1·2·3

Tres Estrellas Rice flour for *atole* drink

Capi Edible fat

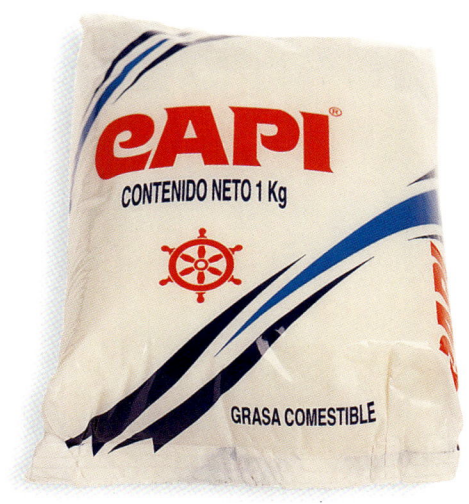

El Labrador — Sugar loaf

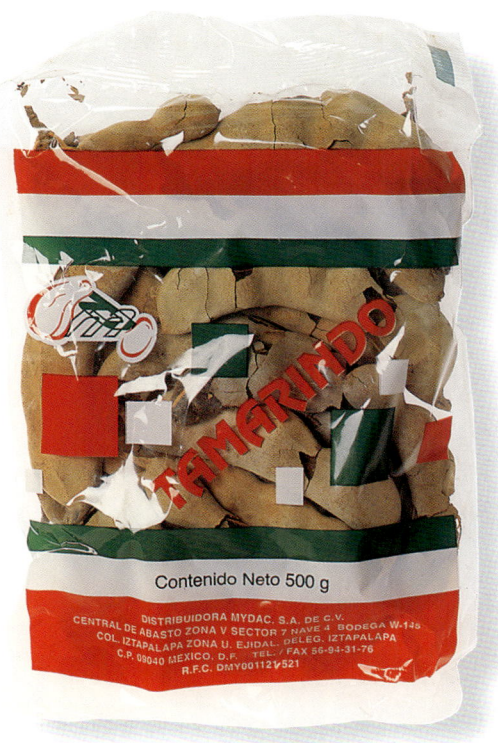

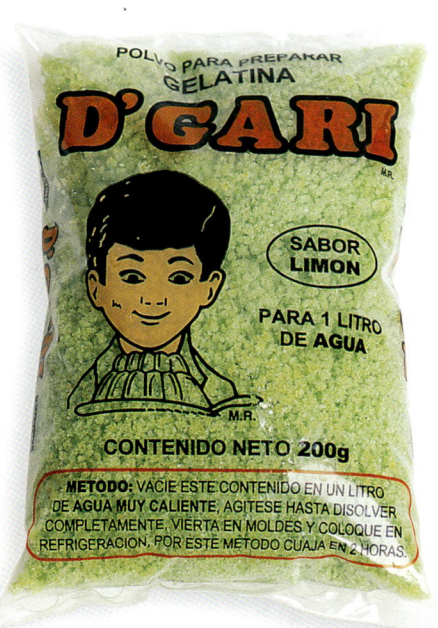

D'Gari — Jello powder

Rexal — Baking powder

Elefante — Sea salt

La Fina — Table salt

Barrilito · Vinegar

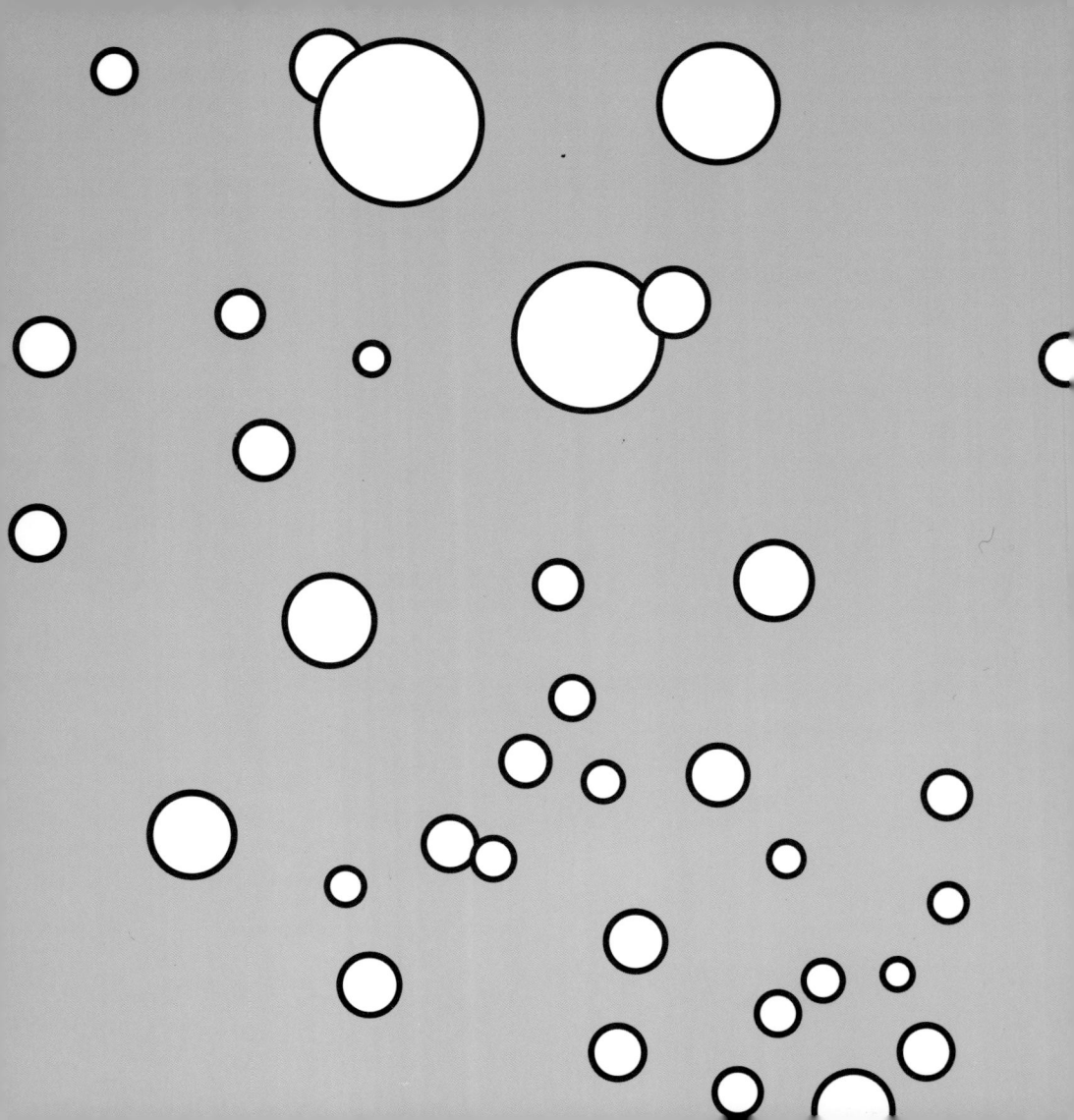

Roma Detergent

Tepeyac Soap for washing clothes

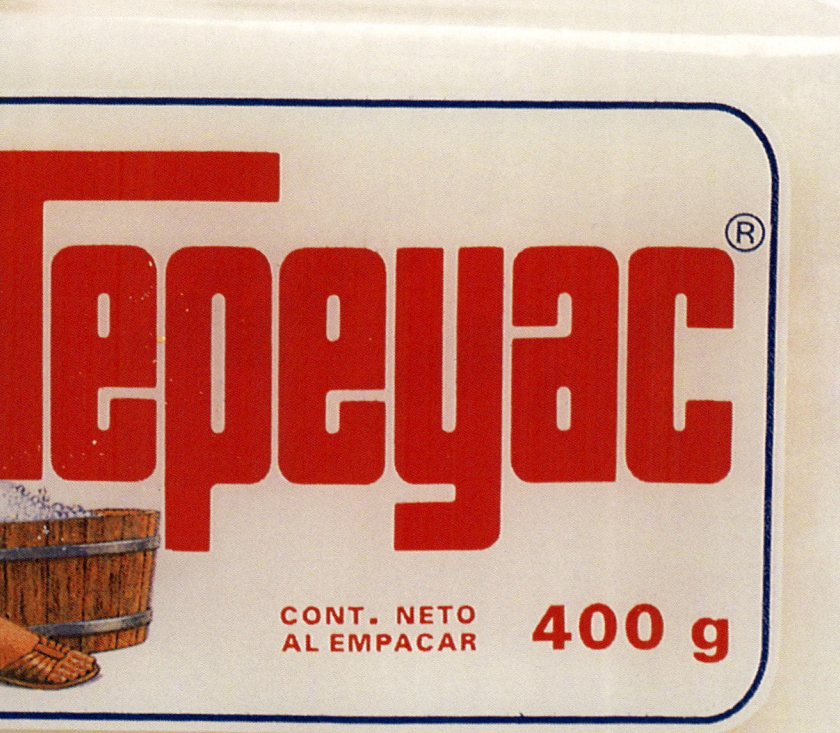

JABÓN ZOT ROSA

Zote — Soap for washing clothes

La Chinita Cleanser

HOT DRINKS

Morelia — Instant hot chocolate

Chocolate Ibarra — Hot-chocolate powder

Moctezuma ▵ Hot-chocolate powder

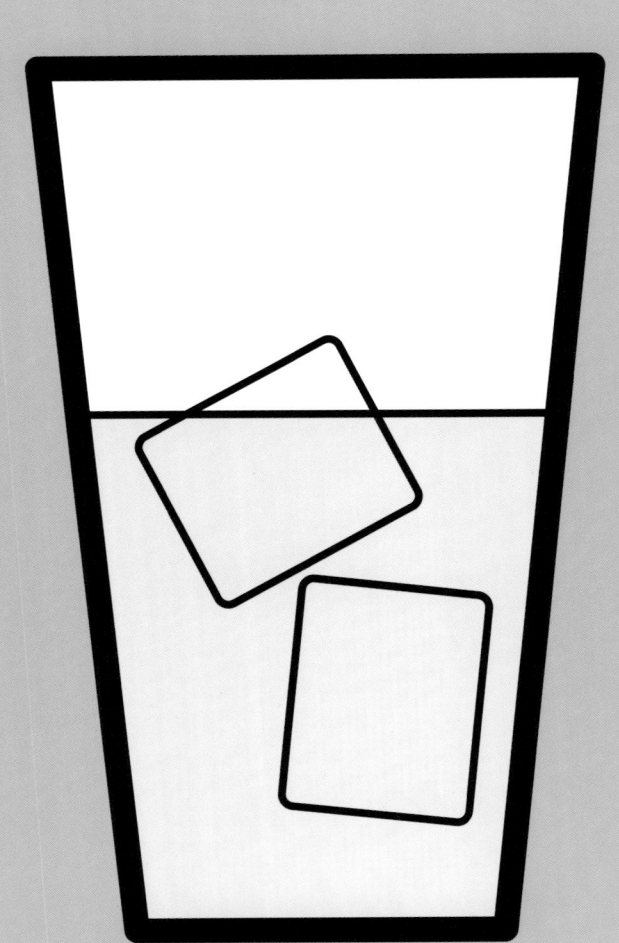

COLD DRINKS

180

Yoli 👕 Soft drink

Boing Fruit Drink

Lulú — Soft drink

Chiva Cola

BEER

Coronita 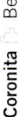 Beer

Sol · Beer

Tecate — Beer

Victoria, León, Pacífico — Beer

SMOKES

202

Maya Matches

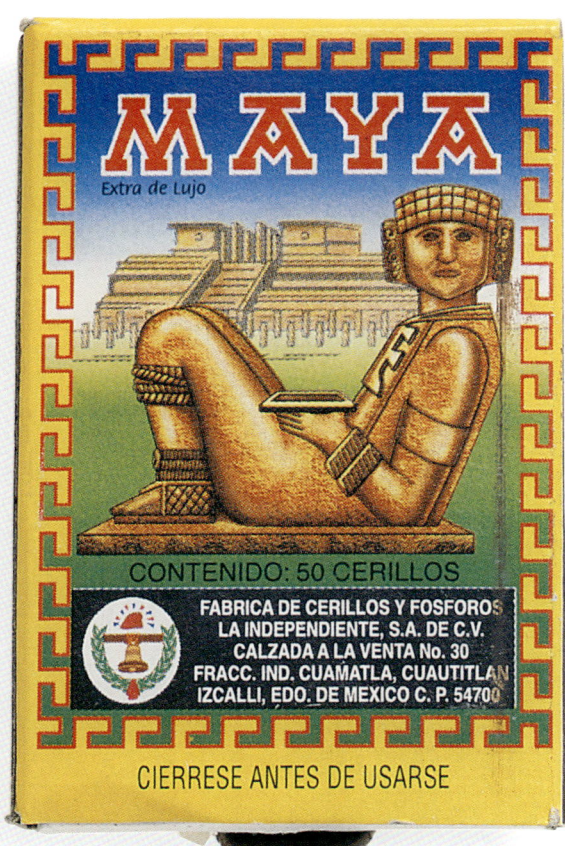

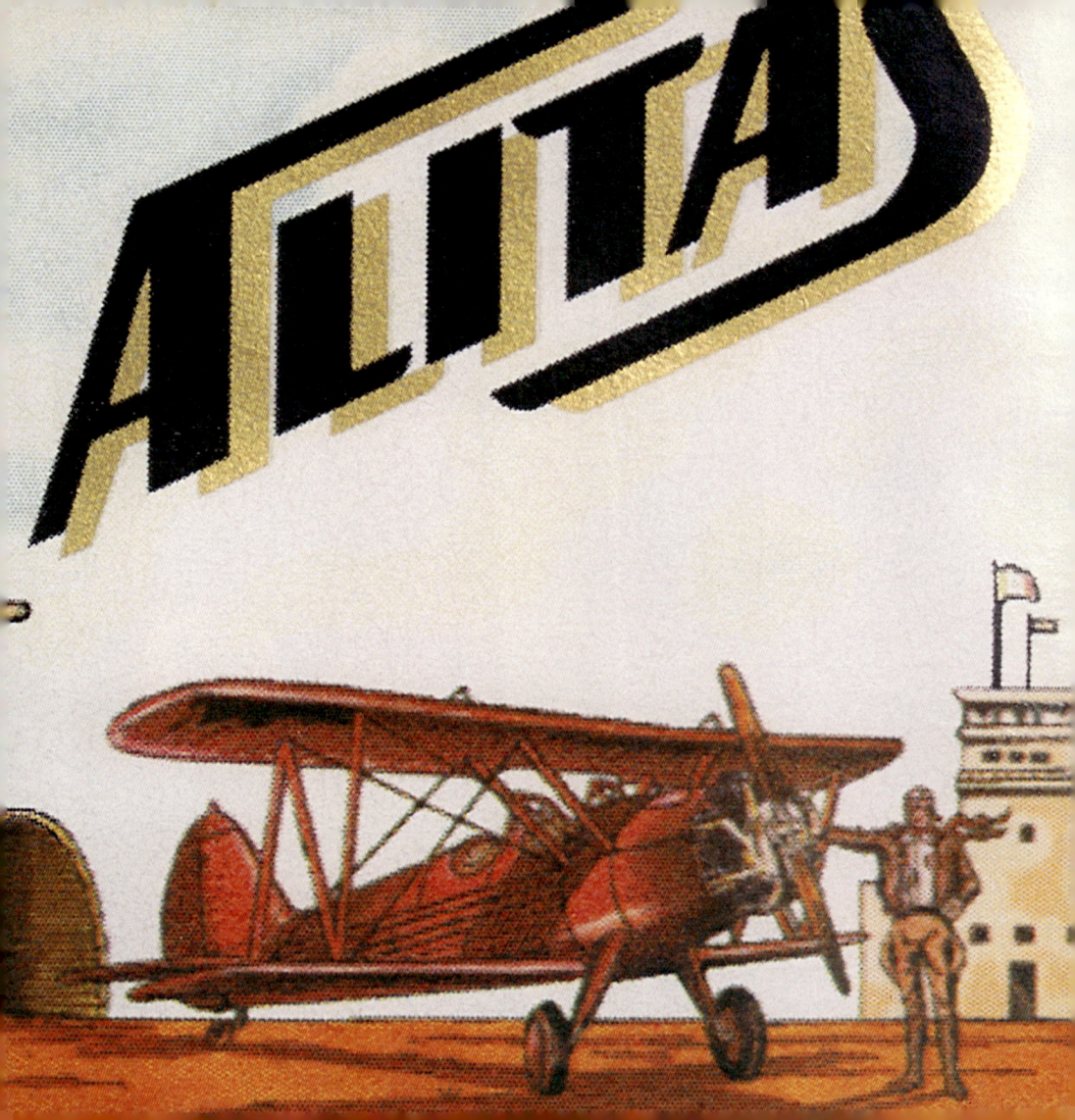

Delicados Cigarettes

ALAS

EXTRA

Alas Cigarettes

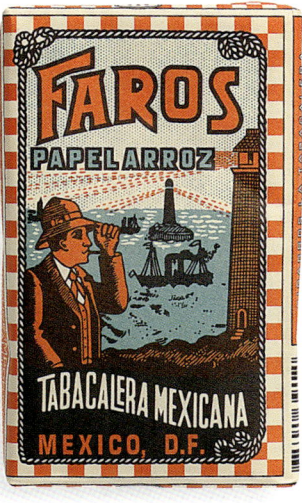

Faros Cigarettes

Tigres ▷ Cigarettes

This book was printed in March 2006.
Trade Gothic LT, Challenge Bold and
Helvetic Compressed fonts were
utilized in the design of this book.